TOTAL ECLIPSE OF THE SUN

Mexico – USA – Canada

PAM HINE

MASCOT® BOOKS

www.amplifypublishing.com

For more information, please contact:
Mascot Books, an imprint of Amplify Publishing Group
620 Herndon Parkway, Suite 220, Herndon, VA 20170
info@amplifypublishing.com

First Edition

CPSIA Code: PREG0123A
ISBN-13: 978-1-63755-210-0
LCCN: 2022915736

Printed in China

CONTENTS

In that great journey of the stars through space
About the mighty, all-directing Sun,
The pallid, faithful Moon, has been the one
Companion of the Earth. Her tender face,
Pale with the swift, keen purpose of that race,
Which at Time's natal hour was first begun,
Shines ever on her lover as they run
And lights his orbit with her silvery smile.
Sometimes such passionate love doth in her rise,
Down from her beaten path she softly slips,
And with her mantle veils the Sun's bold eyes,
Then in the gloaming finds her lover's lips.
While far and near the men our world call wise
See only that the Sun is in eclipse.

—Ella Wheeler Wilcox, 1850-1919

Ella Wheeler Wilcox's poem beautifully illustrates the contrasting ways in which we humans make sense of an eclipse. She interprets it as either "passionate love" between the Moon and the Sun, or "only" that the Sun is in eclipse.

From one perspective, it is a deeply moving event that affects our emotions at a primal level. From another, it's a fascinating intellectual observation of the movement of stars, planets, and moons through space. And in ancient times, many of those who witnessed a total eclipse would often herald the event as a sign of the gods' displeasure or as the precursor of catastrophe.

Regardless of what interpretation we have individually, we all find ourselves captivated whenever the Sun is eclipsed by the Moon.

Within a 6-month period, during 2023 and 2024, we will be treated to 2 momentous eclipses on the American continents:

- an annular eclipse on October 14, 2023 (see page 21)

- a total solar eclipse on April 8, 2024 (see page 9)

Millions of people will experience one (or both) of these events. Around 1 billion are living within the areas of partial eclipse.

WHAT IS AN ECLIPSE?

An eclipse of the Sun (a solar eclipse) occurs when the Moon passes directly between the Earth and the Sun. The Moon's shadow is then cast onto the surface of the Earth.

A solar eclipse is only possible at a New Moon when, of course, the Moon is invisible to us—the side of the Moon facing toward us here on Earth is in shadow.

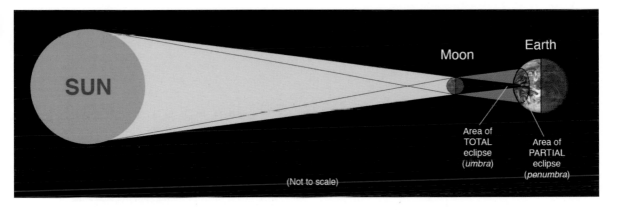

We are all aware of the motion of the Sun across the sky. Day after day, it rises in the East and sets in the West. Many of us are much less aware of the motion of the Moon. In fact, the Moon follows a very similar path across the sky to that of the Sun, but it moves slightly slower, taking on average just under 25 hours from moonrise to moonrise. These apparent movements of the Sun and the Moon across the sky are mainly due to the rotation of the Earth.

Although we may not notice the Moon very often during the daylight hours, it is present in the daytime sky for, on average, half of the time. Once every 29 days or so, at the time of the New Moon (when the Moon is invisible to us), the Sun overtakes the Moon, usually passing above or below it. When the Sun passes directly behind the Moon, we can experience an eclipse of the Sun.

The Earth orbits the Sun on a plane called the *ecliptic plane*. The Moon orbits Earth on the *lunar orbit plane*. These 2 planes are set at around 5 degrees to each other. It is only at the intersection of these 2 planes that it is possible to have an eclipse. This fact helps explain why *total solar eclipses* are relatively rare. (On average, total solar eclipses occur every 18 months.)

The central area of dark shadow from which a total eclipse is seen is called the *umbra*. The much larger area of shadow surrounding the umbra is called the *penumbra*, and this is where a partial eclipse is seen. Within the penumbra, the amount of shadow reduces gradually from the edge of the umbra to the outside edge, indicating the extent to which the Moon covers the Sun.

For around 1¼ hours preceding the total or annular eclipse, the Sun slowly edges behind the Moon during the *partial eclipse* phase. The area of Sun eclipsed by the Moon is termed the percentage of *obscuration*.

During the events described in this book, hundreds of millions of people will be able to witness a partial eclipse throughout most of the Americas (see the global maps on pages 9 and 22).

Make sure to view it all safely.
(See the instructions on the inside front cover and on the arms of the eclipse glasses.)

You MUST use special solar viewers whenever you look at the Sun. (The ONLY time it is safe to observe the Sun without special viewers is when it is COMPLETELY covered by the Moon during TOTALITY.)

When observing the Sun and Moon from here on Earth, they appear as 2 discs in the sky, very similar in size to each other (each being about ½ a degree of the *celestial sphere*). In reality, the diameter of the Sun is 400 times that of the Moon, and the Sun is around 400 times further away from Earth. This amazing coincidence creates the illusion that they are of similar size.

But the Moon's orbit around Earth follows an elliptical path, and the Earth orbits the Sun following an elliptical path, so sometimes the Moon is closer to Earth, and sometimes it's further away. The apparent size of the Moon varies from 10% smaller to 7% larger than that of the Sun.

TOTAL ECLIPSE
(Moon is large enough to completely mask the Sun)

Total Solar Eclipse

Viewed from Earth, it appears as though the (invisible) New Moon obscures, or eclipses, the Sun. It is only when the Moon is closer to us that it is able to obscure the Sun completely, therefore giving us the possibility of a total solar eclipse. This type of eclipse will occur across North America on **April 8, 2024**.

Annular Eclipse

If the apparent size of the Moon is too small to cover the Sun, but it is in line with the Earth, then we can have an annular eclipse (technically a form of partial eclipse). The next annular eclipse will take place on **October 14, 2023**. Only those within the path of annularity in North, Central, and South America will be able to view the "ring of fire" (see also page 21). Safe viewers will be needed throughout this event.

ANNULAR ECLIPSE
(Moon is too small to completely mask the Sun)

PARTIAL ECLIPSE

PARTIAL ECLIPSE
(Moon is not in direct line with the Sun)

There will be a Partial Eclipse visible across a vast area of the globe to either side of the Path of Totality or Path of Annularity (see global maps on pps 9 & 22). A Partial Eclipse will also precede and also follow Totality and Annularity.

LUNAR ECLIPSE

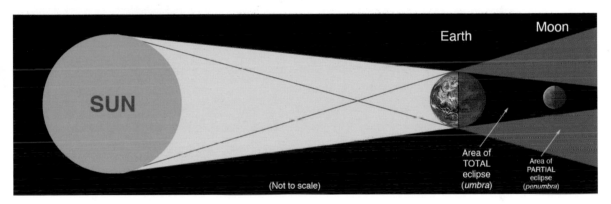

Earth   Moon

SUN

(Not to scale)

Area of
TOTAL
eclipse
(*umbra*)

Area of
PARTIAL
eclipse
(*penumbra*)

A Total Lunar Eclipse (eclipse of the Moon) happens at the time of FULL MOON when the Sun and Moon are on opposite sides of the Earth. The Earth casts a much larger shadow than the Moon and everyone on the dark side of the Earth can see a lunar eclipse. During the total lunar eclipse, the moon takes on a reddish tint.

The *Path of Totality* is created when the shadow of the Moon (the umbra) falls onto the Earth. During this total eclipse in 2024, the Moon's shadow traces a corridor of about 125 miles (200 km) wide, beginning its journey way out in the Pacific Ocean. It makes landfall on the west coast of Mexico, then continues in a northeasterly direction across the US, touching 15 States, from Texas to Maine. It finally crosses the border into New Brunswick in Canada and ends in the North Atlantic Ocean off the east coast of Newfoundland. Scientists will use this rare opportunity to undertake research activities to gain further insights into the Sun's mysterious magnetic field and its effect on Earth's atmosphere. It will be a rare chance to observe the outer atmosphere of the Sun (the corona), which is usually obscured by the brilliance of the Sun itself, attempting to understand the corona's magnetism and thermal structure.

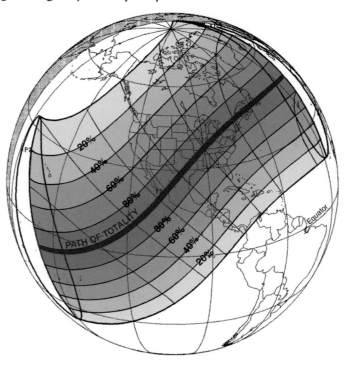

9

These scientists will be trying to answer important questions such as:

- Why is the corona far hotter than the actual surface of the Sun?

- What role does the corona play in the coronal mass ejections that buffet Earth's atmosphere and disrupt sensitive technologies?

- Understand more about the ionosphere (Earth's upper atmosphere) which is home to many low-Earth orbit satellites as well as communications signals, such as radio waves and signals for GPS systems.

You need to locate yourselves within the Path of Totality, as this is the only area where you will be able to witness the total solar eclipse.

It cannot be over-emphasized how essential it is to be WITHIN THE PATH (see page 11). The experience is well worth all the effort to get there. It's a day that people will remember for the rest of their lives.

People in the path of totality will see a spectacular phenomenon—probably the most awe-inspiring light show in the world (weather permitting, of course).

Wearing our special eclipse glasses, we will observe what looks like a small bite taken out of the right side of the Sun. (This is the only evidence we have that the Moon is up

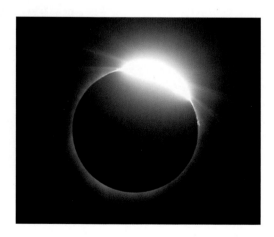

in the sky at that moment, sitting alongside the Sun.) Remember, as we watch this sequence of events (as viewed from Earth), both the Sun and Moon continue to move across the sky, the Sun traveling slightly faster, passing behind the Moon and effectively overtaking it. The duration of the whole event takes around 2½ hours, with totality itself lasting around 4 minutes on the center line.

From the start of the partial eclipse phase, over the following hour or so, the bite continues to grow in size and light levels start to reduce. It will become darker more rapidly as the crescent Sun diminishes to a fine line. It may feel as if a bad storm is approaching, but with no sign of dark clouds, it can feel rather disconcerting. An eerie silence can descend as animals and birds behave as if it is nightfall. We humans are not unaffected by the changes—people often begin to speak more quietly, almost whispering.

As the umbra approaches, we should see a darkening of the horizon toward the west. The crescent Sun reduces to one brilliant spot on the outer edge, known as the *diamond ring* effect. There should also be a fine string of bright beads, called *Baily's Beads*, after the astronomer Francis Baily, who first identified such an effect in 1836. These beads of light are the last rays of sunlight passing through the valleys on the edge of the Moon!

When the last beads have gone, and the Moon has <u>completely obscured</u> the Sun, there will be gasps of wonder from your companions.

It will finally be safe to look at the Sun with the naked eye.

We will experience the awesome and extraordinary sight of a large black disc in the sky, where the Sun should be, and around it the Sun's magnificent *corona*, extending several diameters out into space, like a giant cosmic flower. Many people have called this the Eye of God.

At the edge of the black hole will be a thin red line of the Sun's *chromosphere* (literally sphere of color) along with occasional red *prominences* protruding out from the black hole's perimeter. Stars and planets may become visible in the middle of the day. Don't forget to check out the horizon too. It can appear as if there's a beautiful 360 degree sunset all around. It's a colorful indication that you are standing inside the umbra, the Moon's shadow cone.

As Baily's Beads begin to reappear, the magical phase of totality is over, and it is time to replace your eye protection. If you're near groups of people, there will almost certainly be celebrations, often accompanied by spontaneous applause, cheering, shouting, and lots of hugs. Your relationship with the Sun will never be the same again.

"WHERE AND WHEN IS THE NEXT ONE?" (2025-2030)

Date	Visible in these areas
August 12, 2026	Arctic, Greenland, Iceland, Spain
August 2, 2027	N Africa, Egypt, Saudi Arabia, Yemen, Somalia
July 22, 2028	Australia, New Zealand
November 25, 2030	Southern Africa, Australia

Would you like to be among crowds? In a party atmosphere? In the city? High on a mountaintop? Out in nature in a National Park? Somewhere out on the water in a boat? It's worth some thought beforehand as different locations and atmospheres will provide different experiences.

If you'd like to watch for the shadow as it approaches from the southwest (and moves northeast), it helps to be away from trees and tall buildings.

A few days beforehand, check the local weather reports—the average cloud cover on the list of cities in the path of totality is calculated from historic data (see pages 17 and 18). Because of the changing seasons throughout North America in April, it may be a beautiful sunny day or complete cloud cover or raining! It might be worth making a journey to an area with better weather prospects.

The duration of actual totality varies along the path. On the center line, it can be as long as 4 minutes and 29 seconds in Mexico, reducing to 2 minutes and 53 seconds in Newfoundland. The duration also varies across the width of the path. Because the Moon's shadow is elliptical in shape, the outer edges cast a shadow for a shorter length of time.

This illustration of the shadow shows the importance of choosing the right location with respect to the distance from the center line of the path. The umbra is elliptical in shape as it moves across the Earth's surface, tracing a path around 125 miles (200 km) wide. So, the closer your chosen location is to the center line means the longer that you will stay in the shadow and the longer time you will have to experience the magic of totality.

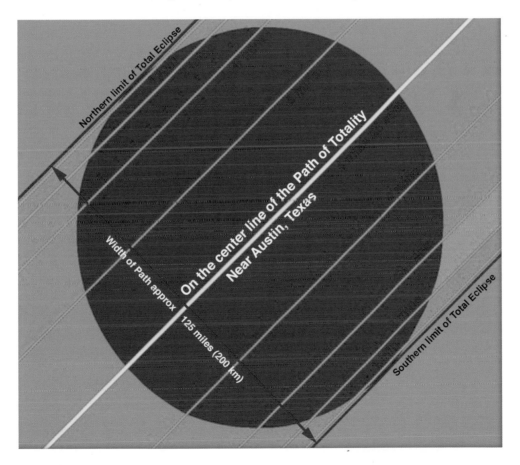

CITIES AND TOWNS ON THE PATH OF TOTALITY

USA

	State	Time Zone	Start of partial eclipse	Start of Totality	Duration of Totality (City center)	End of partial eclipse	Chance of clear or partly cloudy skies
Austin	Texas	CDT	12:17pm	1:36pm	1min 56secs	2:28pm	66%
Fort Worth	Texas	CDT	12:22pm	1:40pm	2mins 39secs	3:01pm	66%
Dallas	Texas	CDT	12:23pm	1:40pm	3mins 42secs	3:02pm	65%
Little Rock	Arkansas	CDT	12:33pm	1:51pm	2mins 29secs	3:11pm	61%
Poplar Bluff	Missouri	CDT	12:39pm	1:56pm	4mins 9secs	3:15pm	58%
Carbondale	Illinois	CDT	12:42pm	1:59pm	4mins 8secs	3:18pm	56%
Paducah	Kentucky	CDT	12:42pm	2:00pm	1min 41secs	3:18pm	57%
Evansville	Indiana	CDT	12:45pm	2:02pm	3mins 4secs	3:20pm	54%
Indianapolis	Indiana	EDT	1:50pm	3:06pm	3mins 49secs	4:23pm	51%
Toledo	Ohio	EDT	1:56pm	3:12pm	1min 47secs	4:26pm	47%
Cleveland	Ohio	EDT	1:59pm	3:13pm	3mins 49secs	4:28pm	47%
Erie	Pennsylvania	EDT	2:02pm	3:16pm	3mins 42secs	4:30pm	46%
Buffalo	New York	EDT	2:04pm	3:18pm	3mins 45secs	4:32pm	43%
Rochester	New York	EDT	2:06pm	3:20pm	3mins 40secs	4:33pm	46%
Burlington	Vermont	EDT	2:14pm	3:26pm	3mins 16secs	4:37pm	45%
Presque Isle	Maine	EDT	2:22pm	3:32pm	2mins 49secs	4:40pm	40%

MEXICO

	State	Time Zone	Start of partial eclipse	Start of Totality	Duration of Totality (City center)	End of partial eclipse	Chance of clear or partly cloudy skies
Mazatlán	Sinaloa	MDT	10:51am	12:07pm	4mins 17secs	1:32pm	57%
Durango	Durango	CDT	11:55am	1:12pm	3mins 47secs	2:36pm	57%
Torreón	Coahuila	CDT	11:59am	1:16pm	4Mins 12 secs	2:41pm	63%

CANADA

Canada	State	Time Zone	Start of partial eclipse	Start of Totality	Duration of Totality (City center)	End of partial eclipse	Chance of clear or partly cloudy skies
Hamilton	Ontario	EDT	2:03pm	3:18pm	1min 52secs	4:31pm	45%
Kingston	Ontario	EDT	2:09pm	3:22pm	3mins 1 sec	4:34pm	43%
Montreal	Quebec	EDT	2:14pm	3:26pm	1min 15secs	4:36pm	42%
Grand Falls-Windsor	Newfoundland	NDT	4:06pm	5:12pm	1min 53secs	6:16pm	34%

On April 8, 2024, almost everyone in North and Central America will experience a **partial eclipse** of the Sun. The maximum proportion of the Sun obscured by the Moon, and consequently the reduction in daylight, depends on how far you are from the path of totality. Los Angeles will experience around 49% maximum coverage of the face of the Sun, while New Yorkers will see around 90% obscured.

(See the global map on page 9.)

A FEW DIMENSIONS

To give you some idea of the scale of the real distances between the Sun, Earth, and Moon, imagine that you are in a football stadium such as the one pictured here.

- In the near goalpost is a **giant beach ball (the SUN)** about 3 feet (90 cm) in diameter.
- In the far goal is a **pea (the EARTH).**
- A **peppercorn (the MOON)** is circling the pea at a distance of about 1 foot (30 cm).

In reality, the actual sizes are a lot bigger:

Sun diameter: 865,000 mls (1,392,000 km)

Earth diameter: 7,920 mls (12,740 km)

Moon diameter: 2,160 mls (3,480 km)

Earth is on average 93,000,000 mls (150,000,000 km) from the Sun.

The Moon is on average 238,860 mls (384,400 km) from Earth.

The ratios of the diameters of Sun:Earth:Moon are around 400:4:1.

PHOTOGRAPHY

Serious eclipse watchers recommend that if this is your first eclipse, just watch and enjoy the whole experience. Recording an eclipse in photographs is an art, and unless you know what you're doing, results can be very disappointing. You might end up with no worthwhile images, entirely missing the main event, when you could have been absorbing the whole experience. There will be loads of fantastic images published by the experts after the event. However, you could easily set up a camera to record the event.

IN NORTH, CENTRAL, AND SOUTH AMERICA

An annular eclipse happens when the Moon is positioned directly in front of the Sun when viewed from Earth, but the moon is not close enough to Earth to appear big enough to obscure the Sun completely. This type of eclipse is often referred to as a *ring of fire solar eclipse* because the outer perimeter of the Sun is still shining brightly.

On October 14, 2023, the maximum obscuration of the Sun will be about 95%, so there will be an obvious change in the level of daylight. However, people within the path of annularity will not experience the extraordinary magic of a total solar eclipse.

We will have to wait almost 6 months, until April 8, 2024, for totality!

The *Path of Annularity* of this eclipse of October 14, 2023, begins its journey across the Earth's surface out in the Pacific Ocean, betwixt the coasts of Alaska and Canada.

It makes its first landfall in Oregon, going on to travel southeastward to touch nine US States before departing the US from Texas.

On entering the Gulf of Mexico and crossing the Caribbean, it lands briefly in Mexico, then goes on to touch Guatemala, Belize, Nicaragua, Panama, Colombia, and Brazil, before dashing off to end its journey across Earth in the southern Atlantic Ocean.

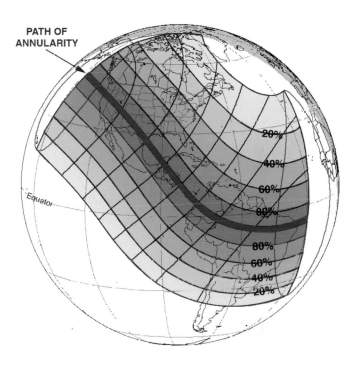

As this global image shows, almost all of North America and much of Central and South America will experience a partial eclipse.

Throughout the whole duration of an annular or a partial eclipse, we will all need to use suitable eye protection when looking at the Sun.

Throughout history and around the world, people have worshipped the Sun as a deity and as the source of light and warmth (indeed of life) here on Earth. The Sun often held a leading place in mythologies. In the past, people took any apparent threat to the Sun very seriously. It must have been terrifying to them to watch as some invisible creature took a bite out of the Sun and slowly consumed it completely. In certain cultures, people often interpreted this event as a sign of the gods' displeasure or fore-telling some great disaster, such as war, pestilence, or the death of kings.

In many cultures, people invented a variety of stories to explain eclipses, such as a god consuming the Sun, or a mythical creature devouring the Sun, such as an evil spirit toad, a dragon, a giant bird, a sky wolf, or a giant bear. People would do whatever they could to scare these invisible creatures away—they shouted, beat drums, pinched their children to make them cry, and beat their dogs to make them howl. It always worked—the Sun has always returned!

MAKE NOTES HERE TO REMEMBER THIS VERY SPECIAL DAY.

Where were you?

Who was with you?

How was the weather?

How did this event make you feel?

JARGON BUSTER

Diamond Ring

The effect seen when the Sun is almost entirely covered by the Moon at the start and end of Totality.

Chromosphere

(sphere of color)
The lower atmosphere of the Sun. It appears as a thin red crescent for a few seconds at the beginning and end of totality.

Baily's Beads

Immediately before & after Totality, the last rays of Sun appear through the mountains of the Moon.

Elliptical Plane

The imaginary plane containing the Earth's orbit around the Sun, or that containing the Moon's orbit around the Earth.

Corona

The outermost regions of the Sun's atmosphere. During totality, the visible area appears as a halo around the Sun.

Celestial Sphere

An imaginary dome of the sky with the observer at its center.

Penumbra

The area of partial shadow of the Moon where only part of the light from the Sun is blocked out. An observer in the penumbra sees a partial eclipse of the Sun.

Partial Eclipse

An observer sees the Sun reduced to a crescent shape as it is partially obscured by the Moon. The observer is located within the penumbra (partial shadow).

Totality

The time when the Sun is completely covered by the Moon.

Umbra

The part of the Moon's shadow where all of the light from the Sun is blocked out. An observer in the umbra sees a Total Eclipse of the Sun.

Prominences

These are flame-colored projections of hot ionized gas rising from the surface of the Sun. We should see them during totality.

The Path of Totality

The route traced by the umbra as it travels across the face of the Earth.

Obscuration

The percentage area of the Sun obscured by the Moon.

- It will take just over 1½ hours for the Moon's umbra to fly across North America from the Pacific Ocean to the Atlantic—around 3,500 miles.

- When watching from Earth, the Sun and Moon each take around 2 minutes to travel their own diameter across our skies. (The Sun is moving slightly faster.)

- It takes around 2½ hours for the Sun to pass behind the Moon throughout a total solar eclipse.

- If you hold your fist up to the sky at arm's length, the width of your fist is a rough approximation of 10 degrees of the celestial sphere. (Apparently, the ratio of a person's arm length to the width of their fist is approximately the same!)

- From "1st contact" of the Sun appearing to touch the Moon during a solar eclipse, to the moment the "4th contact" occurs, the Sun and Moon will appear to have traveled across approximately 40 degrees of the celestial sphere.

- In around 600 million years or so, total solar eclipses will be a thing of the past! The Moon is slowly moving away from the Earth at a rate of 1½ inches (4 cm) each year. This means that at some point it will be too far from Earth to appear big enough to obscure the Sun completely.

BIBLIOGRAPHY

Eclipse map/figure/table/predictions courtesy of Fred Espenak, NASA/Goddard Space Flight Center

Espenak, Fred. "NASA Eclipse Website". Aug 27, 2016. eclipse.gsfc.nasa.gov. Date accessed November 17, 2021.

USEFUL WEBSITES:

Mr Eclipse – Fred Espenak

http://www.eclipsewise.com/solar/SEprime/2001-2100/SE2024Apr08Tprime.html

Time And Date

https://www.timeanddate.com/eclipse/solar/2024-april-8

Xavier Jubier's interactive map

http://xjubier.free.fr/en/site_pages/solar_eclipses/TSE_2024_GoogleMapFull.html

Historic weather information

http://weatherspark.com

iStock Photos

p. 3. Family group wearing eclipse glasses. *LeoPatrizi*

pp. 7 and 14. Totality. *Jorge Villalba*

p. 11. Series of Totality. *TheRoff97*

p. 11. Diamond ring. *yenwen*

p. 12. Baily's Beads. *oversnap*

p. 19. Football stadium. *sArhangel*

p. 23. Myths and legends. *sinopics*

Thanks also to Global Mapping, Nick Ricketts, Hannah Smith, Sue Gross, Richard Hine, my MOAI (you know who you are!), and all at Mascot Books.

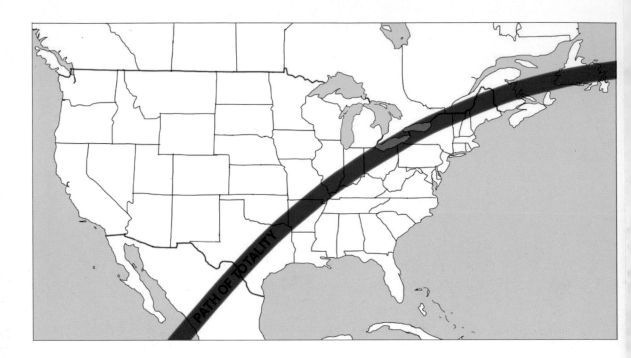

The attached foldout map is marked with timings at 5-minute intervals along the path of totality. Each label has details (in local times) of:

- the time the partial eclipse begins at that location
- the time that totality begins (in red)
- the duration of totality